The Power a
(Biochemical

Llaila Afrika

- ❖ How to Nourish Melanin
- ❖ The Difference Between Black and White People
- ❖ How it Causes Emotional Behaviors and Thoughts
- ❖ How it is Being Destroyed
- ❖ How to Protect It

Other books by the author:
        African Holistic Health
        Vitamins and Minerals from A to Z
        Raising Black Children
        The Gullah
        Nutricide
        Holistic Self Diagnosis
        Pher Ankh

Llaila has a line of specialty disease remedy Supplements. To order Supplements, Telephone Health Consultations, Health and Science classes, Lectures, DVD's, and CDs contact:

Llaila Afrika
P.O Box 501274
Indianapolis, Indiana 46250

Email: llailaafrika@juno.com
Online classes: llailaafrika.com
Official Website: llailaafrika.com
Telephone Number: 317-216-8088

Holistic Therapies and Education Center
P.O Box 501274
Indianapolis, Indiana 46250
Copyright 2014
ISBN 978-0-9896906-2-1

A Charles' Child Production
Melanie D. Stevenson

Cover Design
Jamon O. Lewis

All Illustrations
Llaila O. Afrika

Disclaimer

The author disclaims any liability caused by the direct or indirect use or misuse of the information in this book. None of the information is meant to take the place of an accredited license medical health practitioner.

Manufactured and Printed in the United States of America

Table of Content

Foreword

The subject Melanin is straightforward and simple to explain. Melanin is the biochemical substance that drives physical, mental, emotional, and spiritual life. What makes the subject difficult to explain is the awkward primitive Latin language of chemistry and biology. When Latin is not used it makes information appear unscientific and unintelligent. The challenge in writing about melanin is using social every day language and the avoidance of science Latin Jargon.

The lectures and books on melanin are very Latin jargon centered. Most Black people that read about melanin say, "what good is it?, it does not help you to be free". They of course are correct. A Black person that is unaware of Melanin is manipulated and controlled by their unawareness. Black people need to know the basics about melanin and how to nourish melanin so they can get some positive use from it. Black people are controlled by their failure to realize that the Black race is a Melanin Dominant race. The key to controlling Black People is to reduce their Blackness (Melanin usage and knowledge). This will reduce their ability to be Black, which will directly affect their ability to be human and seek what is humanly theirs- FREEDOM. Black people are a race nourishing themselves as if they are Caucasians. Since the Caucasians have the least amount of Melanin, then Black people that eat as if they are Caucasians are giving themselves the least amount of nourishment to their Melanin. This causes them to utilize their Melanin the least. Therefore, Black people are nutritionally against their own Melanin and themselves. In this writing I have briefly attempted to reveal Melanin's properties and how the human body bio-chemically uses and nourishes it.

The Black race's under-education, dys-education (dysfunctional) and mis-education about Melanin merely reflect one particle of their white domination and Post Traumatic Slavery Trauma. Black people must "know thyself" is to know yourself. To know yourself is to know Melanin.

About Pineal and Melanin

The scientific literature has very few objective facts and many science fairytale stories called theory. All theories are subjective unproven and unscientific. The objective facts are that Black people are a race that has the highest amount of the biochemical pigment called Melanin. Melanin is a civilizing chemical, reproduces itself, a free radical protector, has a sweet fruity smell, can be transformed in the blood, concentrates nerve and brain information, neutralizes, oxidizes (breakdown), converts substances, reduces (builds) another substance and is unchanged by radiation or a high temperature. Melanin is inside and outside the body. The more melanin a race has the more humane and civilized the race. Science myths (theory) have clouded and avoided the true information about Melanin and the Pineal Gland that secretes Melanin. In science myths the Pineal Gland is listed as non-functioning and has no purpose.

Melanin is the vital chemical that makes life itself. It is usually brown to Black in color. Melanin is a flexible like plastic, water, gas, and metal that can resemble a cloud, gas, wood, metal, or liquid. It takes on many forms without losing its structure. It is heat resistant, can endure temperature of 1225°F, has a pleasant odor, resists x-ray diffraction, resist strong acids and alkaline, etc. Melanin is found in the environment, springs, lakes, soil, plants, atmosphere, and animals.

The pituitary gland, which is a gland in the middle of the brain that secretes growth hormones, has been erroneously called the master gland. However, the Pineal Gland, which is also in the middle of the brain, is the true master gland. The Pineal Gland secretes Melanin, which regulates the pituitary gland and all bodily functions and other glands (cycles, circadian rhythms) in the body. Melanin is a biochemical that has a high molecular weight. This means it has many functions and links like a chain that contains carbon-nitrogen, ether, saturated carbon-carbon, unsaturated carbon-carbon, organometallic, peroxides and is essentially a Black carbon hydroamine. These various links allows melanin to be a polymer (takes many forms) and polycyclic (many rhythms) and have a unique property and flexible chemical ability.

Melanin is made by cells called Melanocytes. Inside the Melanocytes are smaller organs (organelle) called Melanosome, which make melanin. Inside the Endoplasmic Reticulum the Melanosomes are made. In other words, small cells build bigger cells, bigger cells build larger cells and the end product of all the building from a small element particle into a larger element particle is the chemical Melanin.

Melanin essentially produces itself with substance such as copper. Melanin absorbs all types of energy such as sunlight, electromagnetic, music heard by the human ear and sounds that the human ear cannot hear, phone waves, radio waves, radar, computer radiation, x-ray, cosmic rays, ultraviolet rays, heat waves, microwaves, etc. Melanin uses the energy in the total environment such as water energy, earth, moon, sun, galaxy, cycles of planets, cycles of minerals etc. On the molecule level the melanin particles called electrons, protons, neutrons, and solatons rearrange their orbits. This is called resonance. In other words, the melanin particles vibrate and rearrange themselves to fill the weak (low) energy sites. Resonance causes a particle to move, this movement causes a small gap (low energy) site and the other particles rearrange themselves to (double bond shift) fill the gap.

Melanin is the natural chemical that makes Black people's skin Black. It is present in Black people's body, skin, cells, nerves, brain, muscles, bones, reproductive and digestive systems in a higher amount than all other races.

Melanin is a biological active substance of various size cells. It is made of nutrients such as indoles, histamines,

3

phenylalanine, catecholamine's (norepinephrine, epinephrine, dopamine, etc.) and the amino acid tyrosine. Melanin is made of various attached parts called chains which are linked to unsaturated carbon-carbon, saturated carbon-carbon, carbon-nitrogen, organometallic, ether, peroxides and quinine, which are brown to black in color. Chemicals such as Flavin, Pteridines, Flavonoid, Napthquinone, Polycyclic Quinone, Anthraquinone, Phenoxazones, convert into Melanin (polymerize or co-polymerize).

The color of Melanin appears as Black because it is absorbing all colors. Once the color enters the melanin it cannot escape. Melanin is a type of concentrated color. It is similar to a cellular Black Hole in outer space. The human eye only sees colors that are reflected away from the object. Consequently, you see black because black is not absorbed but reflected away from the cell. Black is a pigment (non-color) that makes carbon Black in appearance.

Melanin

Melanin is an organic dark carbon chemical pigment substance. Melanin gives Black people's eyes (the iris) a brown color and gives the dark color to their hair, skin and Substantia Nigra of the brain. This is an area of the brain where the cerebrum connects to the pons.

Melanin is secreted by the Pineal Gland. The Pineal Gland has a pine cone shape, is reddish in color and the stem of it is approximately half an inch in length and the head of it is slightly smaller than a green pea and the total weight is about two grams. The Pineal Gland consists of nerve cells similar to those in the retina nerve of the eye. The retina nerve is inside the eyeball and receives light stimulation, which it transmits to the brain. The Pineal Gland is found inside a fluid filled space inside the middle of the brain called the third ventricle and it is attached inside the ventricle.

The Pineal stimulates the growth of the nervous system's cerebrum (thinking brain part called gray matter). Therefore, an under-stimulated Pineal in childhood can decrease thinking ability in adulthood. The Melanin content of the nerves is highest in Black People. Melanin allows nerve messages to travel fast without resistance causing super conductivity. Black people have the highest amount of Melanin on the upper part of the spinal cord (brain stem). This is where the lower part of the brain connects to the spinal cord. The Melanin concentrates in the brains Locus Ceruleus in the 4th ventricle and Reticular Formation (Black dots). This causes the brain to store more information and function at a high level. The function of the body's Melanin and the brain's Melanin is reduced by steroids (cortical adrenals), as well as stimulants (synthetic speeds, cocaine, caffeine), anti-depressants (anti-cholinergics) and unnatural light.

The Pineal gland helps to regulate the hypothalamus gland's release factors that stimulate the Pituitary to secrete the Follicle Stimulating Hormone (FSH) and aides the Luteinizing Hormone (LH). FSH stimulates the ovaries and the movement of sperm and LH stimulates egg and sperm production. Consequently, a Black person with an under-stimulated Pineal or one that uses drugs or eats processed (junk) foods will harm the Pineal and have emotional, mental and reproductive problems.

The processed foods and drugs impact the Pineal. The Pineal regulates insulin levels, the adrenal cortex and the Adrenocorticotropic Hormone (ACTH). ACTH is secreted by the Pituitary and is essential for physical, emotional and mental growth and development. In other words, a Black person's level of thinking (emotional and mental growth) needed to solve their race's problems cannot be achieved with a low amount of Melanin caused by an under-stimulated Pineal.

The melanin secreted by the Pineal aides the posterior pituitary's effect on the hormone Oxytocin. Oxytocin stimulates uterine contraction during the birth of a baby. Oxytocin is similar to morphine; it is one of the main hormones that cause people to bond to each other (mother and child, man and woman, friends with friends). Black people with inadequate Melanin levels will have problems in relationships with each other and getting along together (unity, forming positive groups). They will constantly destroy each other. They will lack the energy to sustain unity because natural bodily made sugars from organic whole foods regulated by melanin cannot be utilized. This is the result of inadequate insulin. The Pancreas insulin hormone takes the natural sugars (not white sugar etc.) out of the blood and puts it into tissue. The Melanin's inability to regulate insulin causes mood swings, irritability and low energy. The liver's ability to make bile is potentiated by Melanin. The Pineal's Melanin hormones act on the liver and if inadequate, this will decrease the gall bladder's alkaline digestive bile fluid secretion. This results in unstable energy and moods. The Pineal stimulates the nervous system and the brain. Consequently, the brain's function is compromised and the type of thinking needed to create ideas to solve emotional, mental and spiritual problems is not sufficient.

The Pineal Glands ability to secrete Melanin is sunlight dependent. The Sun radiates full color spectrums of light. Full spectrum light striking the eyes' retina nerve stimulates the Pineal Gland to secrete Melanin. Melanin turns into serotonin hormone and melatonin hormone. Black people that spend vast amounts of time indoors and out of sunlight or are exposed to the visual pollution of artificial light, sunglasses, television, reflected light of concrete, bricks, highways and buildings, window glass, contact lens, monitors, television screens and fluorescent lights, cell phones, ear phones and in cars tend to have mild depression. A polluted environment, public drinking water, noise pollution, negative moods or social situations, acidic bodily conditions, lack of exercise, processed foods, synthetic drugs and computers decrease melanin production and the stimulation of the Pineal. This visual pollution can result in mild depression because the Pineal Gland is deprived of sunlight and exposed to negative stimulation.

When children are deprived of sunlight it can cause physical and emotional problems. Children who spend many hours indoor (school buildings), playing video games, viewing television and using smart phones, artificial light are deprived of sunlight. Sunlight deprivation results in an under-stimulated Pineal causing the child to grow up and become an adult with decreased reproductive function (decrease gonad weight). In adult animals that have no reproductive organ (gonad) due to castration, the Pineal gets small and the pituitary gets large. Sunlight deprivation decreases the stimulation of the Pineal (almost similar to a castration state condition) which causes one to conclude that the same effect happens with people = Pineal shrinks and pituitary enlargement. The under stimulated Pineal decreases (inhibits) the reproductive organs (gonads) response to Gonadotropins. This causes a decrease weight of the ovaries and testicles. The Pineal gland increases sperm production and helps with female fertility. The Pineal gland increases progesterone hormone (made by the Corpus Luteum).

Melatonin is the hormone that synchronizes the rhythmicity of the body. It helps to control the circadian rhythm. The body has a circadian (cycle) for digesting food. Food ideally should be eaten from 12noon to 7pm, metabolized into the blood from 7pm to 4am and from 4am to noon, the body is cleansing. The Pineal gland secretion of melanin converts into serotonin then melatonin. These hormones need darkness and sunlight to stabilize the rhythm of the body. Sunlight deprivation destabilizes the circadian rhythm causing Seasonal Affective Disorder (SAD). This is associated with abnormal melatonin activity. SAD depression symptoms can be relieved with daily exposure to sunlight and to full spectrum artificial light bulbs and the use of melatonin supplement.

Melatonin supplements along with optimum nutrition, adequate darkness and sunlight are vital. Melatonin release is stimulated by Noradrenaline (norepinephrine), low blood sugar (hypoglycemia), darkness, and Dopamine. The addition of a Melatonin supplement helps Dopamine to calm the emotions and body, helps tryptophan to make serotonin, helps stop antibodies produced by corticosteroids, decreases the growth of certain types of cancers, causes sedation and sleep and decreases the growth of tumors. Melatonin is dependent upon melanin and melanin is dependent upon the Pineal gland and healthy nutrition (whole unprocessed, unchemicalized foods).

Melanin Hormones

The Pineal Gland makes alkaloids. An alkaloid is an organic (means contains carbon) substance that reacts to salts. A salt in chemistry is a chemical compound that is created when an acid and alkaline substance interact. A salt has a positive charged electrical element other than Hydroxyl (commonly called acid) because the alkaloids react to salt that uses them. Salts are used in the body to maintain stable nerves, water balance, regulate the blood volume, are essential for digestion enzymes, hormones, control thickness of the blood (coagulation), pressure inside cells (osmotic), respiratory pigments and acid-base balance. This means that melanin hormones are the primary controller of human life itself. The salts referred to are not table salt chemically known as the poison sodium chloride (chloride is a type of bleach) but the salt of sulfates, phosphates, chloride, carbonates, bicarbonates that are combined with calcium, magnesium and potassium.

The alkaloid serotonin hormone is stimulated by sunlight and full spectrum artificial light. It is secreted into the blood during day light hours. Serotonin connects to other substances (polymerizes) and forms into melatonin. It causes muscles to contract, blood vessels to constrict, regulates blood pressure and has the ability to alter consciousness (improvise other realities). Consequently, the heighten ability to create spontaneously (improvise) is a characteristic of Black people's music (i.e. jazz), dance, acting, dress styles, hair styles, language usage, inventions, science, art and culture. Improvisation is the melanin's ability to think and understand by creating and recreating emotions and thoughts into cyclic neuronet. Emotions and thoughts are bunched together similar to a bunch of grapes. They are an interconnect type spider web. In other words, anger, is connected to love, fear, hate, suspicion etc. You cannot have a single emotion or thought because they come connected to their family (neuronet) of bipolar emotions and thoughts. In the bipolar emotion or thought neuronet the suppression, denial, chauvanizing, or attempts to get rid of the negative pole weakens the use of the bipolar neuronet structure. Caucasians Memorize (2×2=4) and Discover (the sky is blue). Memorizing and Discovering limited brain process is forced upon Black students and people. Memorizing and Discovering is a type of de-polymerizing, un-melinated, synthetic brain(whitening) use of the Black brain. Improvisation is a spontaneous and in the moment of "now" emotional and thought alive thinking.

The alkaloid Melatonin is made by the Pineal Gland. It is secreted into the blood after sunset and reaches its highest levels around 10pm. This alkaloid connects to other substances (polymerizes) and forms melanin. It helps maintain, repair and build brain and nerve tissue. Consequently, it is good for all brain and nerve damage diseases such as Lupus, Alzheimer's, Attention Deficit (a form of Senility), Parkinson, Senility, Memory Loss, Numbness, and Tingling of nerves. Melanin synchronizes body functions, helps to maintain the circadian rhythm, stimulates secretion of milk, and contractions of the uterus, triggers puberty, increases electrical brain wave activity, regulates blood pressure, induces sleep, and stimulates improvising. This gives a positive effect on emotions and creates a feeling of well-being, bonding and joy.

The alkaloids serotonin and melatonin contain a similar structure as other alkaloids. It is the aromatic benzene structure and amino groups that cause the melanin hormones to easily chemically merge with dangerous drugs such as cannibis, Ritalin, cocaine, caffeine, codeine, morphine, mescaline, etc. Black people with their higher melanin content are twice as addicted to synthetic drugs than other races. White Domination which uses the myth of White Supremacy defended by White Racism have created and maintain the physical, mental, emotional, spiritual, global, social, and cultural oppression of Black people. This causes Black people to take drugs, use sex, violence, white sugar, gambling and abuse of each other to emotionally and psychologically escape the oppressive condition. A social condition created by White Domination that causes Black people to socially, emotionally

10

and physically kill themselves is by definition Genocide. The drugs kill and have an effect on the mind, emotions and the physical body of Black people – which means they act upon melanin. The drugs are anti-melanin and against Black culture. The many substances, which are made from melanin, are good in the correct concentration (i.e. alkaloids) and ratios.

Alkaloids are organic nitrogen substances, which are destructive (corrosive) metallic hydroxides (ammonium, carbonates). When the human body makes alkaloids they are not harmful because they are at the correct proportion and concentrations. The Pineal gland makes the unharmful alkaloid serotonin and melatonin. Aside from this, the body makes many toxic chemicals at the correct unharmful concentration such as alcohol, ammonia, bleach (stomachs hydrochloric acid), lye (liver's sulphuric acid, phosphoric acid) etc. However, when synthetic (man-made) chemicals are consumed they bond and incorporate themselves into melanin and generate harmful chemical species that can attack the body at any given moment. This attack can be stimulated by normal body activity as well as negative emotions and social activities. Melanin dominant Black people are highly susceptible to these attacks because they consume synthetic chemicalized junk foods, beverages, drugs and cosmetics. These chemical reactions are related to Melanin dominant Black people, who do not know they have studied Melanin.

Melanin is studied in chemistry, which is the study of Keme (Black melanin particles) called protons, electrons, neutrons and solatons. However, the word melanin is never mentioned in European (Greek) chemistry. Melanin is studied in biology because the purplish (melaninated) brain of the cell called the nucleus is studied. However, the word Melanin is never mentioned. Melanin is the chemical key to life and the brains (nucleus) of all cells. In order for information to be transported to the brain it must be in a liquid form. You see, hear, smell, taste, touch and feel life with your brain (melanin). Therefore, what eyes see, ears hear, tongue tastes, and nose smells' these sensations are converted into a liquid in order to get to the brain. The chemical of conversion or change is melanin. It allows you to be in contact with your inner body and allows your body to be in contact with life itself. The Black race has the highest amount of melanin and has the highest contact with inner life (emotions, spiritual, subconscious, sublime and extrasensory life).

Melanin comes in many varieties and has multiple functional properties. A synthetic drug alters the functional properties of melanin. A synthetic drug that does not act upon melanin is not a drug. A synthetic drug must speed up, slow down, stop or either destroys melanin to be a drug. Any of the body's natural chemical substances or symbiotic virus, bacteria or flora that is outside its range of positive usefulness in the body is un-useful. When it is labeled un-useful it is given names such as free radical, toxin, harmful, parasite, worms and bacterial, viral, fungus and yeast infection. The body's natural chemical substances, bacteria, virus, fungus and yeast can become out of positive usefulness from taking anti-melanin synthetic chemicals, drugs, processed and chemicalized junk foods, negative emotions and behaviors and a polluted physical and social environment. Melanin acts and reacts upon chemicals.

Melanin has a free radical behavior. It will attack harmful synthetic chemicals (i.e. preservatives, food additives, drugs, cocaine, crack, marijuana, Viagra, caffeine) and attach to them in an attempt to transport them elsewhere or neutralize them. The harmful chemical incorporates into the structure of melanin (copolymerize). The harmful chemical and melanin becomes one molecule. This combined melanin and harmful chemical molecule is lodge throughout the body. Emotional stress, disease, physical stressors, junk food, cellphone and computer radiation and drugs can cause the release of this combined melanin and harmful chemical molecule into the blood and into interactions with other chemicals, hormones, minerals and nutrients. The combined melanin and harmful chemical accumulates in the body can cause a toxic overdose, it can breakdown (depolymerize) or reincorporate (polymerize).

Black people's cells have the highest melanin content. Cells build tissue and tissues make organs, organs make organ systems and organ systems constitute the body. The cell has an outer skin (membrane). The cell membrane has a bone type structure called microfilaments, micro tubes (cytoskeleton) and hair like antenna (allele). The antenna senses the liquid environment outside the cell (extra cellular matrix). Melanin is the chemical that converts information outside and inside the cell. Liquid information such as nutrients, red blood cells, white blood cells, fats, enzymes, proteins and hormones need to be constantly read by the melanin. The melanin is the body's computer in a liquid form. It breaks down substances into small units (depolymerizations) and builds small units into larger substances (polymerizations). The information (data) travels on the neurological information highway and is put on a monitoring screen of the brain (Melanin Reticular Formation) and is read by the brain's Frontal Cortex. Information comes from the nerves to the spinal cord to the top of the brain stem to the medulla oblongata to the pods to the cerebellum to the cerebrum. The information highway can have traffic jams, blocks and detours caused by synthetic drugs, junk foods, toxic emotions, feelings and spiritual distortions. The information highway (neurological pathway) problems can create a Black person that is culturally homeless and seeking to serve European culture. Melanin has effects upon Black people's social behavior and has an effect inside the body.

Melanin is highly concentrated in the gastrointestinal tract of the digestive system, as well as the vagina, uterus, penis, sperm storage sac (seminal vesicle), ear (auditory nerves), eye (retina, iris), nervous system, etc. The synthetic drugs enter into the blood stream and are carried all over the body. They are absorbed and interact with melanin in various local areas. These drugs combine (polymerizes) and form different chemicals similar to hybridizing the cells. The cells in local areas resonate or readjust to the synthetic chemical. This creates a type of freak cell similar to genetic modification invaders of the cell. The end result of this biological warfare done by synthetic social drugs and prescription drugs has not been calculated. Black people using illegal (crack) and legal (caffeine, marijuana, alcohol, nicotine) drugs are a walking experiment with unknown results. It is not a matter of **if** Black people will get sick and manifest more physical, mental, emotional and spiritual diseases; it is a matter of **when** they will get sick. Parents that eat junk foods and consume drugs will birth chemically altered children.

The main example of the destruction caused by synthetic drugs is the female's eggs. For example, an adult woman that is pregnant with a girl can damage the girl's eggs. A pregnant woman consuming drugs, marijuana and/or alcohol, has an immune system that can fight the damaging physiological effects. Despite the immune system's attempt to protect itself from the drug the pregnant woman absorb the drug and gets high. The unborn baby girl inside the uterus can fight the effect of the drug. However, the unborn baby girl will get high. Before they are born all unborn girls have all the eggs they will ever have. Unborn baby girls do not grow eggs after they are born; they are born with their lifetime supply. The unborn girl's eggs have no immune system. The synthetic drugs have a direct effect on her eggs. The melanin in the eggs and ovaries connects to the drugs (polymerizes) forming a type of mental, physical and spiritual genetic alteration. Simply put, the pregnant woman is chemically cloning a Black person. The man's sperm has no immune system and the synthetic drugs have a polymerizing effect on it. The man's sperm becomes an alter sperm and in many ways becomes a freak sperm looking for a chemicalized freak egg that will birth freak children searching for White Supremacy to serve and worship. Melanin's healthy impact upon Black people is good and melanin that is altered or damaged has a bad effect upon Black people.

A Black person's emotions and behaviors can be anti-melanin by adopting Caucasian culture as their primary or only culture. African cultural adultery is anti-melanin and is typical of Black people that have no African cultural practices and seek or maintain ways to serve or entertain Caucasians, practice Caucasian holidays and types of relationships and sex. They eat Caucasian processed chemicalized junk foods. Anti-melanin Black people have a lifestyle that allows them to be off their cultural whole organic foods (unprocessed) diet, which means they are off their culture. They are anti-melanin and out of their culture, out of their natural whole food diet and essentially out of their mind and into the Caucasian mind. Negative anti-melanin thoughts, states of consciousness, moods, as well as synthetic drugs cause electrons (minerals with an electrical charge) in melanin to go from a stable state to an excited unstable state. This results in DNA genetic damage, and an abnormal psychological and emotional state and physical illnesses. The negative unstable excited melanin state can cause toxic chemicals (species) that may result in the destruction of melanin and physical diseases. Melanin attaches to synthetic chemicals as a way to grab them and eject them out of the system. However, the continuous consumption of the synthetic chemicals oversaturates and spills the chemical into the blood stream, harming everything with a high melanin content (i.e. brain, sex organs, digestion). In fact, synthetic chemicals eaten get absorbed by the melanin in the digestive system, and then go to the brain and sex organs. The digestive system dumps synthetic chemicals from the intestines into the liver then it circulates in the blood. Synthetic chemicals decrease the liver's immune response, digestive ability, storage of nutrients, and absorption of nutrients such as tryptophan.

The amino acid tryptophan converts into serotonin, which changes into melatonin. During the day serotonin increases while at night melatonin increases. Melatonin aids the growth and repair of tissue by controlling the cells ability to increase its oxygen content, which creates intracellular anti-oxidants. A melanin deficiency causes tissue to destruct or rust (oxidize). Low melanin causes free radicals to damage tissue. Melatonin helps to control prostaglandins. Melatonin allows the body to use sulphur to clean the brain and cells at night. The cerebro-spinal fluid helps to wash out cellular brain waste.

Prostaglandins are a type of fatty acid found in the brain, pancreas, kidney, prostate, uterus, thymus, and lungs. Some prostaglandins are similar to oxytocin and can cause reproductive problems when melatonin is low because of emotional, spiritual, physical or disease stressors. Stressors can activate cortisol and adrenocortical hormone (hydrocortisone), which suppresses the immunity. A suppressed immune system causes many diseases. Diseases have an effect on your mind, mood and state consciousness.

Consciousness is a cultural element. The culture educates a person to have awareness and awareness gives consciousness. Cultures use music, clothes, foods, dance, textbooks, myths, stories, religion, folktales, art and science to create a society. The smallest unit of a society is the individual. The individual is the mirror of a culture. The culture gives a person a belief system and beliefs give a person emotions. A reaction to your emotions is called a feeling. A feeling that lasts a long time is called a mood. In other words, feelings are based upon culture. Consequently a Black person that hears Chinese culture's music will not instantly get up and dance because they have no feelings for that music. There is no cultural connection. A Black person that has been culturally castrated with Post Traumatic Slavery and/or Post Traumatic Colonialism traumas will have a distorted consciousness. Added to this, a junk food (under nutrition) diet will decrease melanin and this causes a decreased consciousness.

A Black person with under nutrition will have physical, social, emotional, and mental problems. They will probably need to take a melatonin supplement.

Melatonin Supplement:
- Helps dopamine to calm the body and brain. Dopamine helps nerves to communicate with each other. It stimulates the hypothalamus and pituitary to release Growth Hormone (GH). Dopamine is required for the ability to sleep, fat gain and loss, sex drive, bone

density, energy, the brain centers immunity and bodily motor control.

- Helps antibodies make cortico steroids a hormone from the adrenal gland that influences digestion and helps the liver make storage sugar (glycogen).
- Helps tryptophan to make serotonin.
- Increases oxygen absorption and life expectancy.
- Helps alkaline the system.
- Helps mental illness.

The body will trigger Melatonin to stabilize Noradrenaline (Norepinephrine), Blood Sugar (Hyper and Hypoglycemia) and Dopamine. Melatonin is released as a reaction to a disease, or an emotional, spiritual, social, or science crisis. A Caucasian science crisis is caused by faulty distorted information that uses European theories (myths).

The problem with understanding melanin is the sciences of Caucasians. Their sciences, contains many, many myths called theories. And their science is White Male Centered. For example, their belief (myth) in Evolution is founded upon the idea that society was formed because men went away from home to hunt for animals. The facts are men, women and children formed society. The hurting of animals in primitive societies is done with nets (large nets for elephants, small nets for rabbits, fish, etc.) and nets are made by men, women and children. Society was not built upon the belief that man went off hunting and women stayed at home. Hunting is a family affair. The foundation of psychology is based upon the behavior of a man called Oedipus. Oedipus is a boy in an ancient Greek fairy tale (myth) that physically lust for sex with his mother. He killed his father and them his mother lusted sexually for her son Oedipus, and married him. The use of this Caucasian psychology myth has distorted a Black person's ability to think healthy and understand.

Psychology's root word is Psyche. Psyche means butterfly. A butterfly has two wings. One wing represents good and the other wing, evil. Consequently, good and evil are always at odds with each other; they are constantly fighting each other. If one moves, the other moves to counteract the other's movement. The wings moving are believed to make the mind think. Added to this belief is the belief that there are 3 divisions to the brain. There is Id the animal brain, Ego that fights to control the animal brain with a human brain and Superego, the mother and father (bipolar nature of society) that fights to control the Ego human brain that is locked into a constant battle with the Id, (animal, sexual, and lust brain). This Caucasian myth is the basis of their psychology and for Black people to use this myth to understand their mind or science presents too many problems. This myth points to the lack of the ability to reach higher levels of thought and physical function.

Caucasians lack the genetic ability to produce (catalyze) significant levels of the heavy molecular weight melanin, called Eumelanin. Eumelanin has its highest content in Black people. Caucasians have a pseudo melanin (Pheomelanin). Their pseudo melanin (pseudo = not a true melanin) can cause their spirit, mind and body to function below an optimum Eumelanin level. This can cause them to have many diseases and disorders of the spirit, mind and body. Black people with Eumelanin can have the ability to manufacture and reproduce melanin by using sunlight, the amino acid tyrosine, the metal copper and perhaps compounds that use oxygen (peroxides) as a free radical. The Free radical use of oxygen allows it to break down substances and food. This is similar to using oxygen to rust (break down) metal. This gives Black people many different (improvisational) ways to utilize the chemical melanin (utilize spirit, mind and body).

Building Melanin

Melanin the dark biological pigment (bio chrome) is built in a specific way (biosynthetic pathway). The amino acid Phenylalanine converts (oxidizes) into the amino acid Tyrosine that then converts (polymerization) into DOPA then into dopaquinone. Tyrosine (oxidizes) makes Dopamine that makes epinephrine (adrenalin) and norepinephrine (noradrenalin). Phenylalanine is found in foods such as raw peanuts, walnuts, lentils, and chickpeas. Tyrosine is found in foods such as beans, oats, wheat, nuts, almonds, avocado, sesame seeds, pumpkin seeds, bananas, and plantains. Tyrosine is found in melanocyte cells inside are granules which are inside vesicles (liquid filled little bumps) which produce Melanosomes (melanin bodies).

Melanosomes leave the melanocytes (polymerizes) and go into the skin, hair, eyes, ears, nerves, sex organs, muscles, adrenal glands, iris choroid, and brain (substantia nigra and locus coeruleus). Melanosomes form Keratinocytes that make barrier cells that protect the top most layers of the skin and iris.

The Eumelanin has the brown and black color and is found in the hair, skin, nipples, eyes, etc. The Pheomelanin has the color red, light red (pink), and yellow. Melanin protects you from UV (ultraviolet) rays. Melanin helps use UV for energy. Melanin covers the cells nucleus similar to a blanket and absorbs the sunlight thus protecting the DNA. Dopaquinone combines with the amino acid Cysteine. Cysteine and Pheomelanin causes red brown freckles on the skin.

Black people are the race with the most melanin (Melanin Dominate). They require at the least 2 to 4 hours of sunlight daily. The sunlight is absorbed by cholesterol then goes to Liver. The Liver emulsifies the sunlight in the oil to make oil soluble Vitamin D (micronage). Black people get less

Vitamin D and sunlight in the cool temperature zone of the earth. If they live above 36 degrees latitude in the Northern Hemisphere or below 36 degrees in the Southern Hemisphere they would need full spectrum light bulbs and about 1000 to 2000 I.U. of Vitamin D daily in the autumn and spring. The closer Black people live to the Equator the better their health and this allows them to get the ideal amount of Vitamin D.

White people have the white gene SLC 24 A5 and the least amount of DMT (dimethyltryptamine). DMT is the spiritual, psychic, and so-called psychedelic compound. Black people have more DMT. White peoples low amount of melanin (pseudo melanin = not true melanin) causes a decrease in the B Vitamins (riboflavin, carotenoids), Vitamin E (tocopherol) and folate (Folic Acid). They have a higher amount of ammonia, salt (sodium), and Sulphur, which can be detected when their hair is wet. Living in the cool temperate zone has very little effect upon them because ammonia gets warm when the weather is cold and gets cold when the weather is hot. If white people get too much sunlight it can cause pathological (disease) occurrence of malignant cancerous tumors of melanoma of the skin.

Melanin is increased (built) on the abdomen (so-called belly) of some pregnant women causing a dark line (linea nigra). Usually during the second trimester (3 months) of pregnancy the linea nigra dark line (melanin) begins above the center of the genital pubic hair and extends up to the navel or to extend above the navel up toward to the breastbone of the chest. During pregnancy melanin darkens the nipples, genitals, and causes moles to darken. Sometimes before the linea nigra appears a pale line (linea alba) is form, which turns dark. The linea nigra melanin helps to make a protective barrier against the effects of ultraviolet sunrays. The biochemical melanin absorbs sunlight, and resists abrasions (damage) to cells. Melanin in the fluid inside the eyeball scatters light beams, which increases the visual field. This means without turning the head a Black person can look straight ahead and can see slightly in back of their shoulder (over 180°angle). No other race has a wider visual field.

How To Measure Melanin

There are many ways to measure melanin's biological, electromagnetic, hormone and chemical activities. The potential (p) for Hydrogen (H), which is called pH, and the Blood Pressure can be used to test Melanin.

The pH is an electrical measurement of the ionic (electrical) element in saliva and urine. The saliva reflects the Autonomic Nervous System's Parasympathetic nerve's actions and reactions. The urine reflects the Autonomics Sympathetic activity.

The test of the pH of saliva and urine samples is done after the person has not eaten or drank fluids (includes water) for two hours, then the samples pH is read with color metric pH paper.

There are levels in which melanin can deviate from normal. Melanin can be insufficient or deficient. A melanin insufficiency can be caused when the supply of melanin cannot meet the demand for melanin. The individual's body can demand excessive amounts of melanin to fight disease and air, water or noise, pollution, radiation from computers and cellphones, junk foods, non-organic food, synthetic drugs, social stress, negative relationships, emotional and spiritual stress. An individual that consumes drugs and eats junk foods and does not use herbal medicine and whole organic foods to defend the body cannot meet this high amount of melanin demanded. A melanin insufficiency is indicated when the urine pH is below or above the normal 6.4 pH of urine. A melanin insufficiency is indicated when the systolic Blood Pressure number of 120 is below or above the normal range. Note, the Blood Pressure norm value of 120/80, in which the top number of 120 is the systolic. The systolic number as well as urine pH is related to melanin's serotonin, the sympathetic nervous system, acidity, the usage of carbohydrates for energy, the left hemisphere of the brain, etc. (see Melanin energy classification).

A melanin deficiency is indicated when the saliva pH is below or above the normal 6.4 saliva pH range. A melanin deficiency is indicated when the diastolic Blood Pressure number is below or above the normal 80 range. Note, the Blood Pressure norm value of 120/80, the number 80 or bottom number is the diastolic. The diastolic number, as well as the saliva pH is related to melanin's melatonin, the parasympathetic nervous system, alkalinity, the usage of raw fats (i.e. nuts, seeds, avocado), the right hemisphere of the brain, etc. (see Melanin Energy Classification).

A melanin deficiency can be caused when there is an adequate supply of melanin and the body's ability to use the melanin is dysfunctional (deficient). The body's metabolic and nerve (neurological) path malfunctions because of a weakened liver, pancreas, kidney, nervous system, respiratory system, reproductive system and immune system. The liver can be damaged due to drugs, sodas, vinegar and alcohol. The pancreas primarily gets damaged due to processed sugars and concentrated sweeteners. The kidneys get damaged from processed sugars and the poison sodium chloride (table salt), high blood pressure and mineral congestion. The other bodily systems can be damaged by hormonal and gland problems that compromise the ability to utilize melanin. The malfunctioning organs, glands, and hormones cannot efficiently use melanin. Therefore, the body is a melanin deficient state.

Races of Humans

Classification of races based upon melanin content inside the body and the skin.

Type	Color	Race
6	Black, Blue/Black, (Highest melanin content)	Africans
5	Black/Brown, Brown	Native Indians (Mexicans, Malaysians)
4	Brown, Red	Native Americans, Japanese
2 and 3	Yellow, mixed, Brown	Orientals
1	White (lowest melanin content)	Caucasians

Health Standards

Medical laboratory normal values, daily recommended allowances of vitamins and minerals, therapeutic dosages of herbs and drugs, baby formulas, disease reaction, human growth and development schedules, brain activity and psychology are all based upon the melanin content of Caucasians. The laboratory and scientific norm values have to be different for each race because each race is biochemically different. The laboratory standards test Black people as if they are white people, and then declare Black people sick or unhealthy because Blacks do not meet White Standards for normal. This is Medical White Racism.

Anatomy
(Difference between the Black and White Races)

	Blacks	Whites
Melanin	✓ Selenium centered Melanin ✓ Higher Molecular Weight (Eumelanin) ✓ High content in body ✓ Increase color absorption in eyes ✓ Increase sound absorption in ears ✓ Acts as polymer ✓ Converts energy ✓ Acts as a computer ✓ Controls cyclical rhythms of all organs ✓ Controls sleep ✓ Controls growth (rate of puberty) ✓ Reacts to gravity (electromagnetic forces) ✓ Highest storage of information ✓ Processes left mind thoughts in right and left hemispheres of brain or middle brain ✓ Processes right mind thoughts in left and right hemispheres of brain or middle	➤ Sulphur centered Melanin ➤ Least amount- causing albinism ➤ Least ability ➤ Lower Molecular weight Pheomelanin (Not a true melanin)

	brain. ✓ Processes largest amount of information in mid brain ✓ High amount of touch intelligence ✓ Can taste the full range of flavor of foods due to melanin cells. ✓ High taste intelligence ✓ Can smell the true aromas has the broadest range of smell intelligence and identification ✓ Highest psyche ability ✓ Absorbs most electromagnetic energy ✓ Highest civilizing ability related to melanin content ✓ Increase memory to memory transfer of stored information ✓ Process most information in corpus colostrum (middle brain). ✓ Evolve highest spirituality due to melanin content	
Skin Melanin (Black Pigmentation)	✓ Allows protection and utilization of the sun's ultraviolet rays ✓ Allows protection	➤ Least of all races, causing white (pink) skin ➤ Reflect colors

	from extreme hot and cold temperatures ✓ Absorbs greatest percentage of colors ✓ Processes more skin cholesterol and sends to the Liver to make Vitamin D (owing to melanin). ✓ Have the most skin pores of any race. ✓ Better cooling ✓ Most skin surface in relationship anatomy.	➢ Poor processor of Vitamin D. ➢ Least pores ➢ Inadequate cooling ➢ Least
Buttock (Stetobygia)	✓ Highest muscular development ✓ Allows extensive hip and thigh movements	➢ Flat, limited mobility
Legs	✓ Longer in proportion to upper body ✓ Allows better movement for walking and running	➢ Short
Tibia (shin bone) **Muscles of Tibia**	✓ Longer ✓ Slender	➢ Shorter ➢ Larger
Blood	✓ When heated (burnt) forms complex pyramids ✓ Allows better storage and transmuting of energy	➢ Less pyramidal

Liver	✓ Slightly large ✓ Allows increased cleansing and energy storage	➢ Slightly smaller
Hair	✓ Least amount of body hair caused by heat insulating effect of melanin ✓ Broadest hair color spectrum	➢ The most hairy of all races
Hair Type	✓ Curly torqued, and brown ✓ Allows quicker transmission and receiving of electrical energy similar to an antenna ✓ Hair shaped like a galaxy (cross section shape)	➢ Flat limp and weak antennas ➢ Least color bands ➢ Hair is closest to fur ➢ Hair has a kidney shape, slightly divided appearance (cross section shape)
Alcohol	✓ Higher amount naturally made by body. Helps to cool body	➢ Lowest amount
Ammonia	✓ Lowest amount naturally made by the body	➢ Highest amount makes then slightly warm when in cold temperatures and problems in hot temperatures ➢ Sun can cause cancer
Eyes	✓ Farthest apart ✓ Allows increased field of vision (peripheral)	➢ Close together, narrow field of vision ➢ Eyes blue, gray,

	✓ Eyes are brown, due to Melanin content ✓ Allows better reception of Sun's color light heat which results in higher stimulation of Pineal and pituitary glands ✓ Iris Absorbs full color, can see the true color of objects. ✓ Orbital opening angular ✓ Wide area seen when looking forward (over 180°) peripheral vision	and green because veins are seen in back of eyes ➢ See paler colors. ➢ Rectangular ➢ Narrow area seen less than 130°
Nose	✓ Broad and flat ✓ Allows angular contour to air columns causing it to vibrate at higher frequency, thus stimulating electromagnetic energy. ✓ Allows wider field of vision for individual eye	➢ Raised chiseled nose bridge blocks field of vision and separates and divides images (sees world divided) limited field of vision.
Women's physique	✓ Buttocks indicate superior muscular movements and counter balance for hips and pregnancy weight. Thighs and Hips wide.	➢ Shape Similar to teenage boy ➢ Hips are wider than shoulders ➢ Poor counter balancing ability
Nerves	✓ High melanin	➢ Least amount

	✓ content in nervous system. ✓ Allows nerve messages to be stored and travel faster and protects against disease.	of melanin of all races.
Jaw	✓ Wider arch and high roof of mouth ✓ Indicates diet high in vegetables.	➢ Narrow, arch and roof of mouth similar to flesh eating animals.
Sulphur	✓ Low amount	➢ High amount, gives offensive smell to hair when wet.
Salt	✓ Low amount	➢ Retains waste fluids in body.
Arms	✓ Longer in proportion to body. ✓ Allow better counter balancing	➢ Short, limit balancing ability
Humerus (upper arm)	✓ Longer	➢ Shorter
Lips	✓ Thick ✓ Allows wider face muscular field and better extraction of juices from plants. Increased sensor action	➢ Thin
Voice	✓ Wider range of speech tones of high and low sounds ✓ Melanin allows melodious and rhythmical speech	➢ Limited range with flat speech tones, tones have no rhythm, lacks melodious sounds.
Ear	✓ Small and stationary	➢ Large ➢ Can move

	✓ Allows better center of sounds. ✓ Inner ear fluid weight and viscosity different	them
Stomach	✓ Has the most bacteria flora (Fungi, Yeast, and Bacteria that live in stomach, entire digestive tract, uterus, vagina, eyes, ears, etc.) ✓ Is specific and unique only to Blacks, have slightly more than 3 pound ✓ Allows food to be broken down (metabolized) at greater nutritional level	➤ No vast variety of flora, limits food metabolism, Tends to have worm population
Vaginal Lips	✓ Larger ✓ Allows tighter seal and increases flora lifespan	➤ Smaller
Vaginal Shaft	✓ Longer ✓ Allows increases bacteria flora lifespan	➤ Short ➤ Decreased bacterial flora
Penis	✓ Length slightly longer	➤ Shorter
Skull	✓ Sagittal counter flat (top of head).	➤ Round
Face height	✓ Low	➤ High
Breast milk	✓ Higher fats ✓ More Alkaline	➤ Higher in Protein ➤ Less Alkaline
Growth and Development Of Child	✓ Fastest	➤ Slowest

Lower nasal margin	✓ Wide base	➢ Sharp
Facial profile	✓ Downward slant	➢ Straight, no slant.
Palate shape	✓ Wide	➢ Narrow
Color	✓ -Eyes darken with age	➢ Extremely rare
Sacral Spot	✓ Birthmark on lower back and/or buttocks	➢ Extremely rare
Breath	✓ Deeper (characteristic of right-minded thinking).	➢ Shallow breath. (Left-minded).
Calcium Intake	✓ Lower Calcium intake (High amount of Vitamin D created by the Liver) stabilizes calcium, reduces need for high intake).	➢ High Calcium intake required
Sterno-clavicular		➢ Found abundantly
Pores of Skin	✓ Widen with age	➢ No change
Muscle	✓ Fast twitch melininated, highly responsive to stimuli, fast action, muscle is light in color	➢ Slow twitch, less responsive, slow in action, muscle is dark in color
Nutrients	✓ Highest nutrient density (most vitamins, minerals and amino acids per square inch)	➢ Least
DMT (Dimethyltr-ytamine)	✓ Increased amount ✓ High spirit and psychic ability	➢ Decrease amount ➢ Low ability
Humerus	✓ Proportion with	➢ Short upper

(Upper Arm)	lower arm	arm
Fibula (lower leg) Calves, tibia, fibula	✓ Proportion with upper leg	➤ Short lower leg
DARC	✓ Duffy Antigen Receptor for Chemokine (DARC) ✓ Protein – skin malaria protection	➤ Narrow similar to young teenage boy ➤ Less Malaria protection
Teeth (cusp)	✓ Identical to Aboriginals, homosapien	➤ Similar to Neanderthal cave people
Laughter	✓ Infants Premotor, Cortical Facial ✓ Muscles develop– 1 to 2 month	➤ Laughter 3 months

Parasympathetic System
(Influenced by the Pineal Glands response to Darkness)
The physical aspects influenced by Melanin's nighttime
production

RIGHT BRAIN: intuitive and special capacities

HYPOTHALAMUS: anterior medial

POSTERIOR PITUITARY GLAND: produces two hormones, controls metabolism, blood pressure, kidney function, smooth muscle action

PINEAL GLAND: is responsive to light, reproductive cycles and pigment

PAROTID GLANDS: helps conserve DNA material, stimulates parasympathetic organs and glands

PARATHYROIDS: Parathyroid hormones releasescalcium from bones

TONSILS: immune system organ, infection warning system

THYMUS: immune system organs

LUNGS: carbon dioxide, oxygen and waste gases exchange from blood

LIVER:energy storage, food processing, detoxification,

GALLBLADDER: bile storage

ADRENAL CORTEX: outer produces several hormones; controls swelling, inflammation,sodium potassium balance: glucocoricoids stimulate production of carbohydrates fom proteins

STOMACH: produces hydrochloric acid for digestion

PANCREAS: stimulated by vagus nerve, produces digestive enzymes: carbohydrate metabolism

SPLEEN: immune system organ, blood regulation

DUODENUM: first part of small intestine: contains opening of pancreatic duct and absorption

SMALL INTESTINE: digestion and absorption

LARGE INTESTINE: B vitamins manufactured, water absorbed

APPENDIX: immune system organ, regulates blood cell production

Sympathetic System

(Influence by Pineal Glands Response to Light)
The physical aspects influenced by Melanin's daytime
production

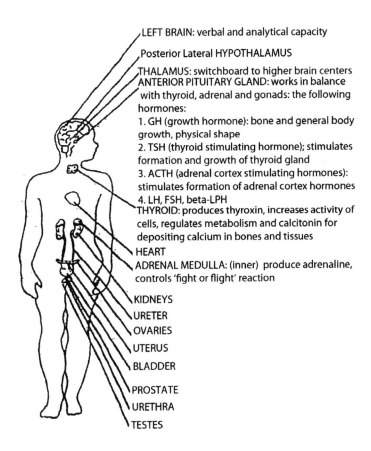

LEFT BRAIN: verbal and analytical capacity

Posterior Lateral HYPOTHALAMUS

THALAMUS: switchboard to higher brain centers
ANTERIOR PITUITARY GLAND: works in balance
with thyroid, adrenal and gonads: the following
hormones:
1. GH (growth hormone): bone and general body
growth, physical shape
2. TSH (thyroid stimulating hormone); stimulates
formation and growth of thyroid gland
3. ACTH (adrenal cortex stimulating hormones):
stimulates formation of adrenal cortex hormones
4. LH, FSH, beta-LPH
THYROID: produces thyroxin, increases activity of
cells, regulates metabolism and calcitonin for
depositing calcium in bones and tissues
HEART
ADRENAL MEDULLA: (inner) produce adrenaline,
controls 'fight or flight' reaction

KIDNEYS
URETER
OVARIES
UTERUS
BLADDER

PROSTATE
URETHRA
TESTES

Arteries	Calcium Metabolism
Capillaries	Cardiovascular System
Ligaments, Connective Tissue	Muscle System
Neuromuscular System	Reproductive System
Skeletal System	Urinary System
Veins	

Human Development –Skin Layers

Each race has a skin layer that is dominate. Black people's dominate skin is the Mesoderm which influences their body, mind, spirit, and chemistry

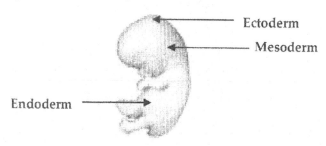

Types of Skin Layers
(Body parts and organs formed from Derma layers of skin)

Mesoderm
> Pineal (melanin), brain, spinal cord, heart, muscle, cartilage, kidney, sex organs, bones, lymph, pancreas
> Mesomorphic People = Africans

Ectoderm
> Bladder, mouth, urethra, lens, cornea, nasal cavity, skin
> Ectomorphic People = Europeans

Endoderm
> Liver, lungs, thyroid
> Endomorphic People = Asians

Note: A disease to an organ or area affects all parts developed from that skin (derm) layer. This is erroneously stated that the disease has spread.

Melanin's Reflection on the Face and Embryo

The mesoderm layer is dominant in Black people. This is where the prefrontal cortex resides and emotions and intellect meet in a balance state.

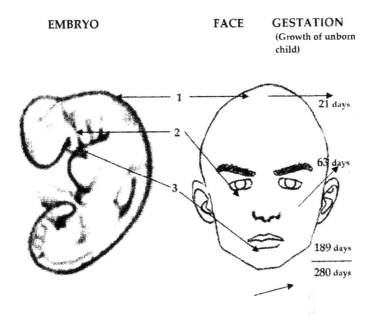

EMBRYO   FACE   GESTATION
(Growth of unborn child)

1
21 days
2
63 days
3
189 days
280 days

1. Ectoderm, 1st Trimester of prenatal growth
2. Mesoderm, 2nd Trimester of prenatal growth
3. Endoderm, 3rd Trimester of prenatal growth

Childhood Development

Fix Reaction Formative Behavior

Child	Adult
Cling to Mother	Hug used for greeting, departing, friendship, affection
Root Nipple (breastfeeding)	Kissing cheeks, lips, derives from rooting nipple
"Koo-ee" sounds reward sound, joy sound	"oo" is derivative of "koo-ee"
a manipulative technique of babies	"oo" it is nice to see you

Childhood, Bonding Before Birth

Fetus Sensors Bond	Complements or Synchronizes
✓ Taste	Movement of Eyes,
✓ Touch	Expressions, Body, Limbs,
✓ Sight	Breathing, Pulse,
✓ Smell	Temperature, and Heart Rate
✓ Hearing	are complimentary and/or synchronized to mothers and father or a man that is bonded to mother physically, emotionally, and spiritually. Synchronization is Melanin dependent.

The human body has one sense, which is the melinated brain. The body has 5 sensors and 1 sense.

Prenatal Emotional Development

Trimester Emotional Development	Behavioral Development	Tendency	Dysfunction Emotion
1st Trimester	Bond/ Reattachment	Clinger/ Avoider Pursuing/ Isolating/ Anger	Scary/ Fear
2nd Trimester	Identity/ Competence	Competitor/ Manipulator Controlling/ Diffusing/ Embarrassment	Anxious
3rd Trimester	Concern/ Intimacy	Caretaker/ Loner Rebel/ Conformist/ Extrovert	Introvert

If the mother is stressed, malnourished or consumes drugs then the child develops these emotions under the heading Dysfunctional Emotion. Children and adults that exhibit these emotions in excess indicate the Trimester of stress and the failure to completely develop a behavior indicated under Behavior Development.

Emotional Vocabulary

Emotions and character logic are similar to each zodiac sign. An emotional vocabulary is essential to bonding to self. Emotions are bipolar (can be bad or good). They develop in a interconnected neuronet similar to a bunch of grapes. All emotions can be used negatively or positively. You cannot have one emotion (fear, anger, etc.) without it being connected to another emotion (love, joy, etc.). Thoughts develop bipolar in a random neuronet similar to a bunch of grapes. We learn to attach thoughts into a logic pattern. Melanin allows us to control and civilize emotions and thoughts in a logic pattern.

Brain

The brain is Melanin dependent and uses Melanin in different forms.

Cerebrum Lobes (Sections)

Lobe	Nutrient/ Hormone
Frontal	Tyrosine, Phenylalanine, Testosterone
Parietal	Acetylcholine, Lecithin, Estrogen
Temporal	GABA, Glutamine, Lipoic Acid, Progesterone
Occipital	Pregnenolone, 5HTP (Tryptophan)

Malnourishment caused by eating junk foods, legal, and illegal drugs and emotional and social stressors which results in inadequate nutrients to the brain and inadequate emotions and thinking. The brain develops thoughts and emotions that are False Positives and see structure (truth) where there is no truth and False Negatives see no structure (no design) where there is design (structure). Dysfunctional thoughts (malnourishment) can cause you to see the truth as lies and lies as the truth. The mind becomes a conflict between thoughts and emotions and is open to control by others (alien culture, white culture).

5 Actions (Variables)
Black people's Relationship Variable
(Influenced by Melanin)

Variables are:
*Social, *Economic, * Political, *Military, *Spiritual

Stimulate movements of Melanin

Outward
Diaphoretic, Expectorant, Colds/Flu

Inward
Liver, pancreas, digestion, seasonings
Upward
Astringents, stimulants

Downward
Diuretics, calmatives, laxatives

Immunity
Blood purifiers, infection, yeast

Psychology

(Difference between Black and White Races)

Black Race (African-Centered Thoughts)	White Race (European- Centered Thoughts)
- Equally uses right and left hemisphere of brain and mid-brain	- Characterized by unhollistic Egotism, illogical use of left and mid-brain, non-spiritual individualism, rationalizations and non-creative.
- Characterized by right-minded spiritual concepts, love, affection, and sharing	- Rationale is based upon conflict between evil subconscious and good conscious. Military logic and predator nature.
- Time exists in "now" and is eternal and cyclic. Future, past, and present are combined	- No present tense of life. Life exists in the past and future: this results in time conflicts, which places no value on present.
- Time is based on the beginning and ending of an event and is composed of the seen and unseen (spiritual, God manifested) causes of an event. Commonly called colored people's time. Time is fixed by the event. For example, the seasons of spring, summer, winter, and fall start according to natures (unseen clock)	- Time is a fixed abstract measurable duration. The seasons start according to a fixed calendar date and not according to nature.
- Thoughts are concept oriented. The meaning of thoughts as well as of words is based on the story	- Thoughts are linear oriented. The meaning of words are fixed and based upon static logic rationales.

(situations) they are used in. For example, the word "bad" can be meant as good, modern intellectual, excellence, or bad. Consequently, this gives rise to statements such as "That's a bad car."	Consequently, a "bad car" means a car unacceptable instead of the Anti-centric meaning of an excellent car.
- Culture is based upon Maat, the family (extended) marriage, ancestors, harmony with nature, children not yet born, spirituality.	-Culture is based upon creating evil to control good and creating good to control evil.
- Communal, Family, and Child Centered.	-Have a pride-type family (similar to animals). -Control of nature -Religions are political systems used to manipulate the powerless -Self centered

- Property owned by society; shared resources	- Property is owned by individuals: No sharing resources.
- Marriages: Predominately polygamous included polygyny and monogamous marriages, many marriage styles and types.	- Monogamous and the practice of polygamous relationships as sexual recreation (illegal gamy)
- Sex is reproductive, regenerational, and spiritual to serve Maat.	- Sex is a physical activity, recreational, and reproductive.
- Individual's value in society is based upon what the individual contributes to society or "you are what you do"	- "You are what you own."
- Economics: Abundantly sharing your goods, talent,	- Economics based upon scarcity, consumerism, or

labor, child rearing, and ancestor's knowledge with society.	the creation of shortages. Thus, only an elite few can gain access to goods, talent, and knowledge, human and natural resources.
- Science is holistic and controlled by Maat.	- Science is belief based (theories) controlled by corporate elite for profit.
- Religion: Monotheism (belief in one God). Do not fear God.	- Belief in many Gods. God created the Devil. You Fear God.
- A person is born to achieve their highest level of humanism	- A person is born into sin.
- Life is based on sense (seen) and nonsense (unseen) and is beyond the power of the mind.	- Life is based upon seen, measured, objective, or abstract knowledge.
- Humans belong to God (no slavery)	- Humans can be owned by man (Full Slavery)
- Psychology based upon moving towards, away from and/or with an ideal and/or emotion.	- Psychology based upon Greek fairy tale Oedipus Myth. Treats emotional illness as if it is mental illness.
- People can be emotionally and/or spiritually ill.	- People can only be physically or mentally (rationally) ill.
- Emotions are bipolar and each emotion is connected to each other (family) in a neuronet similar to a bunch of grapes. Only the ritual for an emotion can be bad. No emotion is bad just the emotional ritual.	- Good emotions are constantly at war with bad emotions (primitive animal feelings).
- Suffers from Post Traumatic Slavery/ Colonialism Stress Disorder	- Suffers from Post Traumatic Cave Life Stress Disorder
- Culture is Male/Female and Child centered based upon	- Culture is Male centered (chauvinistic) Males control

Maat.	history, science, fashion, sports, economics etc.
- Money (i.e. Gold) is a measure of human value (emotional desire)	- Money is wealth and has no human value.
- Thinking is creating and recreating ideas.	- Thinking is memorizing ideas and discovering ideas. Ability to memorize and discover ideas is called intelligence. Amount, measurement, quantity, and/or quotient of intelligence is IQ.
- Values, Wisdom	- Values, Intelligence
- Science advances by improving life	- Science advances from one Theory (belief) to another Theory
Personality Identity on cells membrane called Human Leukocytic Antigen (HLA) = quickly reacts to harmony, symmetry	Slowly reacts
Least amount of humans genes still evolving = mature race	Most amount = immature juvenile race
Forced to Follow the Rule of Law = No privacy, stop and frisk	Allowed to Follow Law Concepts = right to privacy

Melanin Conversions/Melanin Changing Form

Phenylalanine changes to Epinephrine in a progression of steps (Adrenaline)

Phenylalanine ➔ Tyrosine ➔ DOPA (Dihydroxyphenylalamine) ➔ Dopamine ➔ Norepinephrine ➔ (Noradrenaline) ➔ Epinephrine ➔ (Adrenaline)

Tryptophan changes to Melanin

Tryptophan ➔ Serotonin ➔ MSH (Melanin Stimulating Hormone) ➔ Melanocytes ➔ Melanin

Each Melanocyte controls 36 to 40 cells.

Melanin (Pineal) Nutrition

Melanin is the foundation for immunity. It is a free radical scavenger; aids digestion, antioxidant, bones, nerves, cellular, and hormone function. The following are essential for melanin production.

Melanin Hormones: Stimulate	Serotonin/ Sympathetic Stress	Melatonin Parasympathetic Stress
Supplements	Vitamin A B-6 B-12 Calcium Chromium Cobalamin Copper Vitamin C Niacin	d-Alpha Tocopherol B-1 B-2 Magnesium Vanadium Coenzyme Q-10 Vitamin E Vitamin D, Iodine Niacinamide
	Folic Acid Manganese Potassium Riboflavin Silica Selenium Bromelain Iron Chondroitin Sulfate Serotonin	Folic Acid Manganese Pantothenic Acid Phosphorus Pyridoxal 5 Phosphate Riboflavin Choline Zinc Glucosamine Sulfate Melatonin
Herbs	**Combine:**	**Combine**
	Gingko or Gota Kola Damiana, Eyebright Echinacea	Gingko or Gota Kola Chamomile, Echinacea
Amino Acids	Ornithine, Arginine Glutamic Acid SULFER based, Methionine, Cysteine, Taurine	Phenylalanine, Glutamine, Tyrosine Alkaline based: Histidine, Lysine, Glycine
Glandulars	Pancreas, Liver,	Pancreas, Pituitary,

	Brain	Adrenal, Brain

Pineal Nutrition

See How to Measure Melanin to identify the herbs needed for over stimulated serotonin and melatonin over use.

Serotonin/ Sympathetic Stress	Melatonin/ Parasympathetic Stress
Herbs	
Combination: Gingko or Gota Kola Damiana, Eyebright Echinacea, Ashawaganda	Combination: Gingko or Gota Kola Chamomile, Echinacea
Formula	
Saw Palmetto, Black Cohosh, Fo-Ti, Gingko, Lobelia, Eyebright, Burdock, Dandelion Root, Alfalfa	
Foods	
Apricots, Apples, Peaches, Mangoes, Papayas, Star Fruit, Figs, Dates, Bananas, Plantains, Yams, Red Cherries, Blue Grapes, Guava, Broccoli, Spelt, Wild Rice, Blueberries, Collard Greens, Cauliflower, Oranges, Dandelion Greens, Fresh Olives, Raw Peanuts, Strawberries, Currants, Lemons, Turnips, Soybeans.	
Food Classification	
➢ Carbohydrates increase Tryptophan Calmness (melatonin) ➢ Protein Increases Dopamine/Norepinephrine/Tyrosine = Alertness (Serotonin) ➢ Fruits stimulate melanin energy ➢ Vegetables stabilize melanin energy	

Black Peoples Sleep Pattern

Steps	Electric/Magnetic
1. Male principle	Decrease in electricity
Female principle	Increase in magnetism
2. Female principle	Generates magnetism, regenerates
Male principle	Electricity regenerates magnetism
3. Third Eye/Pineal Gland	Generates electricity electromagnetic balanced vibration

Steps	Sensation
1.	Holistic pictures of spirit and physical life.
2.	Feel earth, lunar, solar, and galaxy cycles.
3.	Emotional movement in body, psyche dream trance.
4.	Dream about life in timeless state.
5.	REM-Pineal Gland Vibrations.
6.	Reverse Order (Steps 5,4,3,2,1,) and returns to physical body existence.

African Sleep

Sleep is the Inactivity of the Conscious Mind.
Rest is the Inactivity of the Body

Enter Sleep				Exit Sleep		
Body	**Mind**	**Spirit**	**Black Dot**	**Spirit**	**Mind**	**Body**
Inactive	Voluntary Inactive	Voluntary Active	Enter-Exit	Voluntary Inactive Involuntary Active	Voluntary Active	Active

Cyclic Sleep Sequence

Cycles

Earth	**Lunar**	**Solar**	**Celestial**
Physical	Conscious	Electrical	Magnetic
Inactive	Emotion,	Decreases	Increase
	Inactivity	See Pictures	Floating

Sleep Supplements

Calcium	3,000mg	Magnesium	1,500mg
Vitamin B	100mg	Vitamin b$_6$	300mg
Complex	300mg	(Pyridoxine)	
Vitamin B$_5$		Melatonin	1-3mg
(Pantothenic	As directed	Nutritional	3tbsp
Acid)		Brewers Yeast	
Gaba			

Herbs

Catnip, Passion Flower, Hops, Skullcap, Chamomile, Lady's Slipper, Valerian, Kava Kava

Melanin Energy Classification

These are various chemical and mental processes stimulated by melanin's serotonin and melatonin

The Male Principle (Serotonin)	The Female Principle (Melatonin)
Go	Stop
Sympathetic	Parasympathetic
Acid (fast moving)	Alkaline (slow moving)
Vertical	Horizontal
Destruction	Construction
Catabolic	Anabolic
Anaerobic	Aerobic
Systolic	Diastolic
Melanin insufficient	Melanin deficient
High specific gravity	Low specific gravity
Hemolyzed	Non-Hemolyzed
Action	Relaxation
Future	Past
Left brain (Rationale, Analytical, Realistic)	Right brain (Emotional, Harmonizer, Idealistic)
Lower jaw	Upper jaw
Incisors	Molars
Think 1st, Feel 2nd	Feel 1st, Think 2nd
Contraction	Expansion
Constipation	Diarrhea
Insulin	Glucagon
Carbohydrates	Fats
Urine - waste	Saliva – enzymes
Bright colors	Dark colors
Objective	Subjective
High turbidity	Low turbidity
High pulse	Low pulse
Science	Art
Salt	Sugar
Vagina	Penis
Fight/Flight	Tend/Befriend
Oxidant (oxidize)	Antioxidant
Extroverted	Introverted

Neurons of brain	Glia cells of brain
Acetylcholine	Dopamine
Explicit	Implicit
Enzymes	Receptors

Melanin Hormone Symptoms

Excess stimulation of Serotonin and Melatonin has these symptoms

Serotonin/Sympathetic Stress	Melatonin/Parasympathetic Stress
Dry mouth	Saliva and tear quality increased
Eyes set inward	Eyes set outward: exophthalmos with irritation of cranial nerve III (e.g. form exophthalmic)
Fat energy uses decrease	Fat energy usage increase
Glucose energy increased (insulin)	Glucose energy decreased or increased
WBC decreased	WBC increased
Poor circulation associated with vasoconstriction	Poor circulation associated with decreased pulse pressure
Heart; Kidney; BP problems	Heart problems
Indigestion; ulcers; gallbladder; and bowel problems	Indigestion; ulcers; bowel problems; colitis
Food allergies	Allergies; asthma
Immunity low	Immunity high
Increased thyroid activity	Decreased thyroid activity
Pupil large	Pupil small
Increased sensory perception	Decrease sensory perception
Pulse increased; arrhythmia	Pulse decrease
Respiratory rate decrease or increase; Bronchial dilation; Respiratory depth increased	Respiratory rate decrease or increase; Bronchial constriction
Systolic BP and pulse pressure increased	When standing, failure of systolic BP and pulse pressure
Urine specific gravity high	Urine specific gravity low
Urination decreased	Urination increased

Temperature increased	Temperature decreased

Melanin Hormone Symptoms

Serotonin Sympathetic Stress	Melatonin Parasympathetic Stress
Cold sweat on hands, or cold dry hands	Hands warm and dry
Nervous tension; tremors	Nervous tension; depression; anxiety
Vasoconstriction	Vasodilation
Breathing increased	Breathing decreased
Penis Flaccid	Penis erect
Digestion decreased	Digestion increased
Sexual desire decreased	Sexual desire increased
Sleep decreased (sleepless)	Sleep increased
Aggressive	Passive
Serotonin high	Serotonin low
Melatonin low	Melatonin high
Rheumatism	Arthritis
High energy	Low energy
Colors: red, yellow, orange	Colors: blue, indigo, violet
Decreased blood to the skin	Increased blood to skin
Mania	Depression
Decreased progesterone	Increased progesterone
Testosterone	Estrogen
Increased during day	Increased during day

12 Melanin Clusters (Chakras)

Melanin energy pathways are erroneously labeled Acupuncture meridians. These meridians collide (cluster) in areas that are erroneously labeled Chakras. Only Melanin Dominant Black people have 12 clusters, other races have less. The clusters can be stimulated by hand (massage, acupressure). Bio magnetics and Acupuncture needles. The stimulation can cause wellness and help remedy disease.

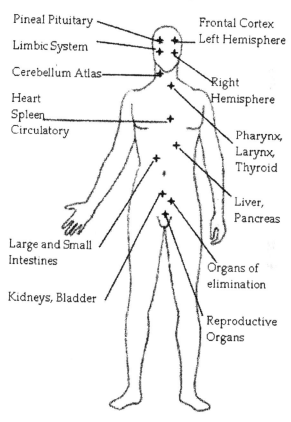

Melanin Colors of Zodiac Holistic Chart

Body	Zodiac	Music Note	Melanin Colors	Zodiac House
Pituitary, Eyes, Cerebrum, Nose	Aires	D	Red	1
Neck, Cerebellum, Throat, Lower Brain, Thyroid	Taurus	E	Yellow	2
Arms, Fallopian tubes, Thymus	Gemini	F	Violet	3
Breast, Veins, Arteries, Stomach	Cancer	B	Green	4
Heart, Thymus, Spleen	Leo	A	Orange	5
Intestine, Liver, Pancreas, Lower Abdomen	Virgo	C	Violet	6
Kidney, Adrenals, Bladder	Libra	D	Yellow	7
Genitals, Gonads, Rectum	Scorpio	E	Red	8
Sciatic Nerve,	Sagittarius	F	Purple	9

Thigh, Hips, Buttocks				
Knees, Teeth, Bones	Capricorn	G	Blue	10
Para-thyroid, lower leg, Colon, Ankles	Aquarius	A	Indigo	11
Feet, Circulation, Pineal	Pisces	B	Indigo	12

Wholistic Chart

This Chart relates melanin clusters to Melanin colors and other Holistic aspects

Melanin Clusters	Glands	Kenetic Word	Colors	Music Note	Number Sound & Shape	Letters Sounds & Shapes
Third Eye	Pineal	IKH	Indigo	B	6	FOX
Crown	Pituitary	Mer	Violet	A	7	GYP
Throat	Thyroid	Sekhem	Blue	G	5	ENW
Heart	Thymus	Kheper	Green	F	4	DMU
Solar Plex	Adrenal, Liver	AB	Yellow	E	3,8	CLU, HQZ
Genital	Ovary, Testicle	Tekh	Orange	D	2,9	BKT, IR
Perineum	No Gland, Colon, Rectum	Seteklit	Red	C	1	AJS

Index

noise pollution, 7, 22
Non-hemolyzed, 52

O

osmotic, 9
ovaries, 6, 7, 14
overdose, 12
Oxidant, 52
oxygen, 15, 17, 18
Oxytocin, 6

P

Pancreas, 6, 47, 58
Penis, 52
pituitary gland, 2
polymer, 3, 25
pregnant, 14, 20
prenatal, 37

R

reproductive problems, 6, 16

S

sex, 10, 15, 16, 17, 19, 36
sleep, 8, 10, 16, 25
sperm production, 6, 7
supplements, 8, 65

synthetic speeds, 5

T

Trimester, 37, 39
tumors, 8, 20

U

ultraviolet, 3, 19, 20, 26
Upper jaw, 52

V

Vagina, 52
virus, 12

X

x-ray, 2, 3

Y

yeast infection, 12

Z

zodiac, 39

Suggested Reading

Afrika, Llaila: African Holistic Health

Barnes, Carol: Melanin, The Chemical key to Black Greatness

Barr, F.E. "Melanin The Organizing Molecule" Medical Hypothesis Vol. 11: 1 March 1983

Boyd, William: Genetics and the Races of Man

Coon, C.S: The Races of Europe

Gan, E.V. Lam K.M. "Oxidizing and Reducing Properties of Melanins" British Journal of Dermatology 8b (25) 1977

Hawley Gessner, G: The Condensed Chemical Dictionary

Kaidbey, Khet: Photoprotection by Melanin A Comparison of Black and Caucasian Skia J. American Acad Dermatol Vol. 1 (34) Sept. 1979

King, Richard: Black Dot (Humanities, Ancestral Blackness, The Black)

Loeb, Stanley: Nursing, 2007 Drug Handbook

McGinness, J. Proctor P: The Importance of the Fact that Melanin is Black J. Theor Biol Vol. 39, pp. 677-678

Morrison, W.L: "What is the Function of Melanin? Arch Dermatol, Vol. 121, pp. 1160-1163 Sept. 1985

Price, Weston: Nutrition and Physical Degeneration

Prodor, P.H. Hilton, J.G., and McGinness J.E: Meeting on Biophysics and Biological Function of Melanins Pharma, 3-4, March 1980

Quay, W.B.: Pineal Chemistry Charles Thomas, Springfield, III 1974

Salazar, M.M. "Significance of the Interaction of Drug-Binding to Melanin Biochemical Pharmacology, Vol. 28 pp. 1181-1187, 1979

Thomas Clayton: Tabers Cyclopedic Medical Dictionary

Vaugh, G.M. "Melatonin in Humans in Pineal Research Review Vol. 2, pp. 141-201 Alan R Liss Inc. New York, NY 1984

Welsing, Francis: The Cress Theory (Racial Confrontation)

Williams, Richard: Textbook of Black-Related Disease

About The Author

Llaila O. Afrika is a Doctor of Naturopathy, Licensed Acupuncturist, Licensed Medical Massage Therapist, Certified Addictionologist (treats all types of addiction), Certified Nutritional Counselor, Medical Astrologist, Spiritual Counselor, Marriage Counselor, Psychotherapist, and worked as a Nurse in the Military. He has treated people of all races for diseases and has lectured in North and South America, Caribbean, Europe, and Africa. He has been leading pioneer in the African Science field and has books such as *African Holistic Health* (been on Essence Magazine Bestseller List for over ten years and is the largest selling book of its kind), *Nutricide*, a book about harmful drugs and foods that Black people consume, *Raising Black Children* (infancy to teenagers), a comprehensive text on parenting, children's behavior, growth, emotions, diseases, and healthy foods, *The Gullah* which is a history book, Holistic Self Diagnosis is a textbook that teaches with illustration, charts, and pictures for skills to diagnose diseases, and *Pher Ankh* teaches how to arrange and decorate the homes, rooms, buildings, and gardens for health, beauty, success, and power, *Vitamins From A to Z* explains how to use supplements, herbs, homeopathic remedies, and food to prevent and cure diseases.